# MANAGEMENT SYSTEMS THEORY – A PRACTICAL APPROACH

## WITH PUBLISHED ARTICLES AND CASE STUDIES

By Issah Yakubu Wurishe

# TABLE OF CONTENTS

INTRODUCTION                                                    3-5

CHAPTER ONE                                                     6-12

DEFINITION OF MANAGEMENT SYSTEMS

CHAPTER TWO                                                     13-16

THE APPLICATION OF MANAGEMENT SYSTEMS

CHAPTER THREE                                                   17-19

EMPIRICAL PRESENTATION OF MANAGEMENT SYSTEMS

CHAPTER FOUR                                                    20-22

IMPACT ASSESSMENT OF MANAGEMENT SYSTEMS

CHAPTER FIVE                                                    23-24

MANAGEMENT SYSTEMS RESOURCES

CHAPTER SIX                                                     25-26

MANAGEMENT SYSTEMS TOOLS

CHAPTER SEVEN                                                   27-39

ORGANIZATIONAL RESTRUCTURING (A CASE STUDY APPROACH)

CHAPTER EIGHT                                                   40-60

PUBLISHED ARTICLES

APLLYING MANAGEMENT SYSTEMS IN THE AGRICULTURE SECTOR IN GHANA

MANAGING THE RICE INDUSTRY FOR SUSTAINABLE FOOD SECURITY AND ECONOMIC
GROWTH IN GHANA

RURAL TO URBAN MIGRATION VERSUS RURAL INDUSTRIALIZATION IN GHANA

BIBLIOGRAPHY                                                    61

# INTRODUCTION

Management Systems is both a theoretical and practical activity, involving the science of planning, organizing, staffing, directing, coordinating, communicating, controlling, motivating, and decision making. These activities are managerial processes which are systematically applied.

Management systems can be defined as the main structures, pillars or poles upon which an organization is built and functions. These structures are apparently necessary for the performance of any managerial function and the successful realization of any managerial goal or target. These structures or pillars include: production; marketing; personnel; finance; communication; research and development; and any other structure deemed appropriate and necessary in an organization.

The importance and value of management systems are immeasurable and cannot be fully quantified. This is because, the benefits begins with the commencement of the organization and flows along with its successes and failures, and seemly addresses the challenges which confronts managerial initiatives, innovations, decision making and strategies. The relevance of management systems is therefore as explained above, and includes the following as well:

Firstly, all organizations work with certain quantities and types of resources which must be used judiciously and sometimes converted into consumable goods and services. According to Howard Barnett (1992), all organizations have five different kinds of resources to work with, and they are; equipments, material, money, people, and time. The basic idea of resource conversion into usable goods and services involves the adoption of these management systems, coupled with the utilization and improvement of managerial functions to achieve the desired goals.

Secondly, modern portfolio theory postulates a trade-off between risk and return. This means that, the establishment of any business entity entails some level of risk; and the quantification of any net return will depend on the efficient functioning of these management systems, such as the finance, marketing, personnel systems, etc. In other words, modern investment management is based on the efficient risk management and this is only achievable through the establishment of a functional management system in the organization.

Thirdly, management systems in some industries are facing acute challenges, leading to low productivity. This is because, the management systems are ill established and therefore continue to fall off from their peak best to low best scenarios; in terms of earnings and capitalization. This problem is peculiar with small firms, which are sometimes managed haphazardly by their owners. Moreover, as they are mostly ran by their owners, these firms do lack the requisite talents and skills to apply modern managerial techniques. Story (1982), explains that, studies conducted in the past proved that firms established by graduates performed much better in terms of turn-over than similar ones established by non-graduates. An enterprise is run as an art of skill acquired through orientation or otherwise, and not as an opportunistic venture.

In addition, bigger firms have their own peculiar setbacks which downplay some of their activities. One of them is, management systems are not applied in a vacuum, instead, they require the application of certain resources and managerial techniques to raise productivity. The antidote for bigger firms in this respect is for them to apply the correct tools of management systems to match adequate resources at the same time. The management systems resources include people, technology, money and assets; and information. Barnett (1992) writes about personnel at work, and according to him, a person is qualified when they have been properly trained to do the job, and are used to doing it. Another setback to bigger firms in applying management systems is their inability to find and use information to make informed decisions. Information as mentioned is gathered

4

through research, which plays an important role in determining the development, sustainability and growth of an organization.

The management systems tools are research data; team work; strategic plan; and timely decisions.  These tools are applied to ensure the judicious use of scarce resources and the ability of the organization to realize its goals and targets.

# CHAPTER ONE

## DEFINITION OF MANAGEMENT SYSTEMS

Many management theorists and founders have defined the term management in different ways. Henry Fayol, the celebrated mining engineer and management theorist wrote; to manage is to forecast, and play to organize, command, coordinate, and control. The definitions of management have differed over time, relative to the environment under which it is described. Many of the definitions basically stipulate and identify the functions of management. The management functions as stipulated in some of the definitions, especially in that of Henry Fayol are: planning; coordinating; controlling; organizing; motivating; directing; communicating; staffing; forecasting, etc. The elements of management, specifically planning, are functionally applied through the management systems.

These functions cannot be performed in a vacuum rather they are channeled through the management systems or structures. These systems or structures includes: personnel; production; marketing; finance; communication; research and development; and any other which may be deemed or found to be necessary.

Etienne et al (1992), defines a system as a set of components linked by relatively organized relationships, in order to fulfill certain functions. Management systems employ and use all the levels of management and link them together to achieve the organizational goals and targets. Without a coordinated process of these systems or departments, no meaningful activity can be undertaken within or outside the organization. The resources of the business are converted into finished goods and services by the management systems. To ensure an efficient and effective management practices and performance in an organization, these systems or department are programmed to work in units and subunits; and in teams and sub teams; led by managers and supervisors, who are sometimes called directors and managers. Thus, management becomes a wholesome and functional system, only when all the activities of the various departments are well coordinated and streamlined to achieve the desired organizational goals.

Personnel System

The personnel system is also referred to as the human resource department. It is established to ensure the steady and smooth direction of human activity in an organization, including welfare and safety. Barnett (1992) writes that, this department has the responsibility for everything that affects a person at workplace, and some employees involve themselves in ensuring their general well-being.

This management system cannot function properly without qualified personnel being employed to do the work. Alan et al (1992) also clarifies the role of the personnel manager as consisting of two parts: firstly, is the supervisory work; and secondly, the direction of the conditions of work of the employees. The personnel system is charged with the key responsibilities of supervising the employees to do their work diligently; ensuring their welfare and safety; and providing adequate training and development programs for them. The main activities in the personnel department can be classified as planning, recruitment, staff development, labor relations, health and safety, reward systems, and welfare.

Production System

In the economics perspective, the production system operates with some factors of production, namely, land, labor, capital and entrepreneurship. In the agricultural perspective, Etienne et al (1992, describes the production system as characterized by a combination of certain types of production factors, including, manpower, land and equipment. In the broader sense of management systems theory, the production system involves the judicious use of all the available resources of the organization to meet its production targets. The production system is the lifeline of the organization, because it is the primary source of revenue generation.

The production system must be effective, efficient, evolving and viable. According to Appleby (1991), the production system involves an effective planning and control of the operations of that section of the organization, devoted to transforming materials (resources) into finished products. The main activities of the production system, among others are; quality assurance, design, creation, standardization, simplification and specialization of the products.


Marketing System

The marketing system begins where production system ends. It mobilizes all the resources available to the organization to ensure profit, survival and growth. The marketing system is an entrepreneurial activity which involves the judicious and speedy disposal of the end products to the final consumer.

Foster (1982) describes marketing system as the systematic application of entrepreneurship to the almost constantly changing conditions of the present age. The British Institute of marketing defines marketing system as the management process responsible for identifying, anticipating, and satisfying customer requirements profitably. Similarly, the American Marketing Association defines it simply as the performance of business activities that direct the flow of goods and services from producer to customer or user.

The basic activities of the marketing department are promotion, product modeling, sales, distribution, and agency. These activities determine the viability and sustainability of the organization.

Finance System

The finance system of any organization is the availability, provision and management of all resources of the entrepreneur or investor. There must be the available funds classified as initial capital investment or seed money, to purchase the production factors needed to kick start the business. The application of the initial capital investment must be subject to the level of risk involved in the business.

According to Dubbins and Witt (1983), it was Harry Markowitz who laid the foundation for the evolution of modern portfolio theory, especially in the aspect of risk quantification, which suggests that, for any given level of risk, the rational investor would choose the maximum expected return; and that for any level of expected return, the rational investor would choose the minimum risk.

The finance system is charged with the greatest responsibility of generating and managing cash inflows, as much as the organization will need. As the organization invests its scarce resources into the operations of the business, it expects to rip big fruits from its investment. The returns on its investment will largely depend on its capitalization from the markets, sales, diversifications, profits, restructuring, and stocks.

The initial financing of a business does not depend only on personal savings and assets, but also includes loans from individuals and institutions, and as well as goods and services bought on credit. The financial system is very important because its key activities pervade the other departments in the organization. Some of the key activities of the financial system are: risk management; profitability, assets and accounts management; cash inflows and outflows resulting from sales and market returns; stocks and share management; and debt management.

## Communication System

The communication system in an organization involves the effective transmission of information from within and outside the establishment to the appropriate quarters, where the information is needed. It embraces the management information systems, administrative work, records keeping, and computer based systems, which includes database management. Mardock and Scutt (1993) also defines communication system as the provision and passing of information and instructions which enable a company or any other employing organization to function efficiently and effectively and employees to be properly informed about developments.

The communication department is the main liaison of the organization to its employees; and between the organization and its external relations. It is a vital, pivotal and reserve point of information seeking for the organization itself and other relations. The main activities of the communication department includes: communiqués, meetings, negotiations, service time schedules, programs and projects, policy advocacy and dissemination, and sometimes training. The communication department stands out to be as important as any other department, since information is the bedrock of any surviving organization.

## Research and Development

Research and Development system is equally an important pole in any organization, since the use of appropriate data stems from research and other investigative services. No organization can prosper well without the use of ideas from scientific research into areas of prime concern within or outside the entity. The research and development department is responsible for collecting, analyzing and introducing various data into the organization. It is also responsible for initiating and introducing various standards of improvements, maintenance, and innovations into the organization; and ensuring that such standards are continuously updated to conform to the dynamics of time and the preferences of end users.

This unique department will link with the other departments in the organization to determine their needs, and systematically provide them with the statistical data they require to perform efficiently and effectively, especially in decision making. For example, research into personnel system will involve process study and analysis of work, work study, method study, work and time measurement, ergonomics, etc. Barnett (1992) emphasizes that, new product development should link closely, not only with the marketing department to ensure that what is being developed closely matches with customers perceived needs, but also with the manufacturing and industrial engineers to help ensure that the product can be made in the most economically affordable way. Similarly, the pricing of a product requires some research to determine the right price for the product, taking into cognizance the interests of both the producer and consumer. Lawrence Miles (1947) categorized the overall value into parts, namely, value in use and value in esteem. These two standards require a research study to determine the best price for the product, based on consumer perception and affordability.

In conclusion, management systems can be quantified to determine which direction the organization is moving towards, whether the organizational growth is progressive, stagnant or retrogressive. In this sense, organizational growth is equal to the cost of management systems divided by profit earned, preferably over a one year period. It can therefore be represented mathematically with the formula: $OG = \dfrac{COMS}{PE}$

Where: OG means Organizational Growth; COMS means Cost of Management Systems; PE stands for profit earned.

Furthermore, management systems is a departmental network, comprising personnel, production, marketing, finance, communication, research and development and any other system, to ensure efficiency and effectiveness in the running of an organization. It organizes the entire organizational hierarchy into a common front, to run corporate activities and functions more prudently in order to achieve the desired objectives and targets. The overall performance targets and long term goals of the organization are achieved through the management systems.  Subsystems or units can be created under the management systems to ensure proper and effective team work within the organization.

The practicability of this management systems theory is more robust in big and well established firms, and some medium scale businesses.  It is also highly recommended for small businesses, if proper management processes and procedures are to be upheld.  It is very possible to see some or all of the management systems being applied in many organizations in the developed world than in the developing countries.  This is so because, in the developing countries many businesses are either medium or small scale enterprises, and as a result they feel reluctant to adopt the appropriate management innovations such as the management systems.  Lack of adequate finances of the businesses accounts for some of these reluctances and failures to adopt management systems in the developing countries.  Secondly, lack of consistent growth rates in some of the developing economies is impacting negatively on the operations of businesses within their domains.

# CHAPTER TWO

## THE APPLICATION OF MANAGEMENT SYSTEMS

Management systems theory is applicable in all spheres of human organizations, though relative to the sort of classification or identification given to it.  All economies of the world face the same economic problem of scarcity of resources relative to our needs and uses. That is why governments are faced with the superlative   challenges of being given the mandate to oversee and consciously direct their economies to meet their pressing needs.  In order to achieve these objectives, the interests and responsibilities of the citizens are classified and assigned to their respective systems for redress to meet the national goals and aspirations.  The government is able to organize and oversee the smooth direction of its management systems through an institutional tax system, which obliges every capable citizen or organization, otherwise any liable non-citizen, to contribute his or her quota from their earnings; and also from the naturally endowed resources of the country.  In addition, economies are dominated by industries and businesses, which are either controlled by the state or are privately owned by individuals or associations.  However, the success of all these economic activities depends on the mode and depth of application of the management systems.  The establishment and application of management systems in government and businesses should not be done haphazardly, but should be applied systematically and judiciously.  It should be done with a certain mode of classification of the processes, resources and tools of management systems.

Methodology

The first method of applying management systems in an organization is the overall management systems category.  The overall management systems method embodies the responsibilities of all the systems in a single person, who is the owner.  The sole owner is his or her own employer, financier, and marketer.  Basically, everything revolves around one person.  He is the 'Jack of all'.  This method can be successful if the owner is innovative, hardworking and takes good decisions.  This method survives when the operator uses scarce resources judiciously, and his sole aim for this kind of business organization is meant for the short to medium term, but is never sustainable over a long period of time.  The economics of scale is never in favor of such businesses.  This method is applied when the manager is also able to keep records and accounts of the business, to ensure that his or her expenditure falls far below his income, because the vice versa is unacceptable for such a small organizational. Since his or her resource base is small coupled with the volatility of the business environment, the entrepreneur must plan over a specific period of time to expand the business.

The second method of management systems is the delegated or mixed type. This method is suitable for partnerships, consortiums, cartels and growing businesses. This method allows for certain management systems to be merged and assigned to a particular department. For example, merging the roles of the marketing department to communication; or taking away some roles of the finance department to the marketing department, if such realignments will augur well for the smooth running of the organization for the short to the medium term. Otherwise, such delegations or mergers are appropriate if the decision to do that is more viable and profitable to the organization. The critical determining factor for this type of management system's application is the availability of resources.

The third type is the structured management systems method, which is suitable for large and complex organizations. These types of organizations can apply either the delegated or structured methods. For the structured management systems method, all departments are established to operate interdependently on their distinctive roles created for them. They perform distinctive roles which are linked together towards the achievement of a common organizational goal.

Challenges

There are many and varied challenges associated with the establishment and operation of management systems in an organization. Each of the management systems have their peculiar problems that they contend with. The main problem associated with personnel system is remuneration. Remuneration is an important concern to entrepreneurs because, it is either determined by the employer or is negotiated between the employer and the employee. Good remuneration can lead to high morale and productivity, whereas the reverse can lead to unstable industrial relations and low productivity.

On the other hand, the major challenge associated with the production system is efficiency and effectiveness. Efficiency relates to the methods adopted by the production system to achieve its production targets. Whereas effectiveness relates to how best production targets can be achieved. The main challenge affecting the marketing system is economic depressions, especially when it leads to low incomes and widespread unemployment among the active population. This challenge can lead to decline in market and sales and low consumption among the populace.

The main challenge, on the other hand, affecting the finance system is the ability to mobilize adequate resources and then being able to use the available resources efficiently. The burden of debt is what must be avoided at all cost, because excessive indebtedness can lead to the liquidation of the organization.

The main challenge of the communication system is when wrong decisions and policies are made and implemented. The main challenge confronting the research and development system is when inaccurate statistical data is used to plan for the organization. It can result into serious conflicts, instability, misdirection and misapplication of scarce resources.

Importance

The application of management systems is very important to the economy, industry and businesses in general. Every economy and business must be organized and managed on the correct paths and principles; except in peasantry situations where simple methods are employed. Considering the economy as a whole, a nation will apply management systems in order to convert its scarce resources into utilities. The establishment of the management systems will involve the creation of ministries and agencies to be assigned with specific functions in the economy. These ministries and agencies will represent the pillars of the economy, organized to work together in a coordinated effort to achieve the common good of the nation, namely, economic growth and development which will contribute to the raising of the standards of living of the people. Similarly, industries and other businesses will equally appreciate the significance of management systems in the achievement of their respective targets and goals of existence.  Therefore, the application of management systems in the economy, industry and business is important because of the solutions it provides to the numerous challenges they are faced with.  Most importantly, management systems provide good marketing management and planning, effective and efficient production and personnel systems.  It also provides an efficient finance and communication systems, and a resourceful research and development system.

# CHAPTER THREE

## EMPIRICAL PRESENTATION OF MANAGEMENT SYSTEMS

There are scores of big, medium and small industries operating around the world. The size of each industry relatively depends on the variety and quality of its products and the supply chains. Typical examples of some of these big, medium and small industries are the rice industry in the indo-china region; wheat and maize industries in America; automobile industries in Japan and USA; Oil industry in the Middle East; Gold and Diamond industries in Africa, etc. These classifications are based on the aggregation of output levels, but not necessarily on competitive production figures. The industries mentioned are among many of their types around the world, operating within the global economy and supply chains. These industries are affected by the trends in the global markets, such as good governance, fluctuations in prices, trade regulations, inflation, exchange rates, industrial relations, etc. Despite these challenges, some of them still grow and prosper, and most of them who are successful are those who have put in place competent and functional management system.

The Growing Rice Industry in Ghana

As far back as 1911 the Ministry of Food and Agriculture had started testing the suitability of rice varieties in the industry. According to Jean (1994) rice cultivation in Ghana started as far back as the seventeenth and eighteenth centuries. The rice industry is predominantly private small scale farmers and then, the bigger organizations which are either owned by the government or are sourced through multi lateral agreements which include agencies under the UNO, NGOs such as GIZ, USAID, etc. Some of the governmental organizations are the Ministry of Food and Agriculture, Irrigation Development Authority, Council for Scientific and Industrial Research. These governmental organizations have structured management systems in place, and they are managed by government appointees. The privately owned enterprises are solely managed by their owners. Management systems are not adequately applied in some of these organizations; in some instances some of them do not even have qualified personnel to support in the decision making process of the organization.  They seasonally employ casual labor during planting and harvesting periods.  Most of these businesses do not have premises to operate from, nor do they have adequate resources to establish management systems such as marketing and finance.  On the other hand, the multi lateral agencies and NGOs who operate in the agriculture sector provide goods and services which directly or indirectly benefits farmers.  These organizations have the mixed or structured management systems in place, to execute their plans and projects.

The lapse in the application of modern managerial practices in the industry is as a result of many factors.  Firstly, many of the farmers are illiterates, about 53.5% of them are either illiterates or do not have formal education or are elementary school drop-outs ; Whereas about 46.5% have either completed secondary or tertiary education.  Secondly, lack of adequate capital to invest in their businesses, to enable them to apply modern managerial systems.  Thirdly, many of the small holder businesses are not registered with the registrar general's department to qualify them for formal and mainstream support from banks and other agencies.  Furthermore, majority of them are peasant farmers who are unable to expand their business beyond their household needs.  For example, in the northern region where rice is mostly grown, about 98.75% of the farmers cultivate only rice   or with other food crops like maize, cassava, yam, etc.  Out of this large proportion of rice farmers, about 60% of them cultivate between 1 acre and 50 acres; and 40% cultivate more than 50 acres.  Also, in terms of investment capital about 71.25% invest between $16 and $400, and only about 25% of the farmers are able to invest over $400 during the farming season.  As a result of the poor capital investments in the agriculture sector, majority of the farmers do not have access to irrigation facilities, therefore they are left at the mercy of weather variance.  In addition to these is the menace of pests and diseases, bush fires, lack of agro processing industries to add value to the commodities, and the lack of common rice varieties.

The challenges facing the bigger organizations which apply either the mixed or structured management systems are mainly, lack of research and development programs and adequate investments capital.


Thus, in developing countries most businesses are small or medium scale and are therefore reluctant to apply management systems due to lack of capital and well trained manpower.

# CHAPTER FOUR

## ASSESSING THE IMPACT OF MANAGEMENT SYSTEMS

As we note in chapter three, the nature of the organizations operating in the rice industry in Ghana can be classified into three types. Firstly, the privately owned businesses which are predominantly small or medium size organizations. These types of businesses rarely apply modern managerial systems. Secondly, the government owned establishments which are well structured with mixed or structured management systems in place. Finally, the NGOs some of which are multi lateral agencies or are non-profit organizations established to support farmers to realize their basic needs.

On the other hand, businesses are mostly classified into three types, namely, sole proprietorship, partnership, and joint-stock company. The organization and management of these types of businesses depends on the application of management systems.

Therefore, the nature and degree of the impact of management systems will depend on the type of business organization that is being dealt with, its operational level on the basis of the type of management systems being applied, and also its business and product potency in terms of marketability, capitalization and profitability.

In order to measure or assess the impact of the application of management systems in an organization, the following factors needs to be considered: productivity; programs and projects being undertaken; forecasting techniques; resource management; technical skills and application; availability of reliable data; availability of assets and investments; and the application of strategies.

High productivity is achievable through good production planning, involving the triangulation of market needs, and motivation through the efficient and adequate supply of inputs and raw materials.

The efficient and effective management of programs and projects indicates the efficiency and effectiveness of management systems.  Projects are drawn from programs and according to Barnett (1992) a project is a one off job consisting of many tasks.  All of these tasks have to be defined clearly.

Another standard for assessing the success of management systems in an organization is the capacity to forecast reliable trends and targets.  According to Barnett (1992), an organization's success depends on how accurately it can forecast the level and timing of demand for its goods or services.  The structures of the organization must be resourced to be able to accomplish forecasting activities.

Resource management is a key factor one has to consider when assessing the impact of management system in an organization.  The resources available to every organization is scarce relative to its needs, hence the relevance of an efficient and effective management systems application.  Therefore the judicious use of scarce resource wholly depends on an efficient and effective management system.

The success of applying management systems in an organization will depend on the application of technical skills, so where an organization has competent technical skills and adequate numbers, then the impact of management systems on the organization will be positive.  Solutions are found from research and available data and this activity should be continuous and based on competent technical skills.

An organization should be viable before people would want to invest in it. Most investors will look for minimum risk and profitable ventures with sound management systems. These guarantees are found in well established organization with structured management systems.

Strategic management is associated with good and efficient application of management systems. Strategic management decisions involve all levels of the organization, especially the top management; and every strategy will be implemented with an action plan drawn up from the various management systems.

CHAPTER FIVE

MANAGEMENT SYSTEMS RESOURCES

Management Systems resources are in different forms, depending on the nature of the organization and the products and services it deals in.  The resources are used to enable the various structures in an organization perform their functions perfectly and within the ambit of the organizational goals and targets.  These resources can be categorized into people, assets and money, and information.

The human factor in any organization is very crucial to its very existence and survival.  This is because, the human resource transcends in all aspects of organizational governance.  It is the people who act as entrepreneurs or labor at the same time.  As a management systems resource, people are used to design and manage plans and resolve issues and conflicts affecting corporate governance.  It is the single resource when missing in an organization can lead to a total standstill and possibly disintegration of the organization.  It is people who perform functions which cannot be performed by machines.

Assets and money are the bane of most organizations.  It is not easy for an organization to be self sufficient in this resource.  The absence of this resource can cripple the organization's ability to fulfill its dreams and aspirations.  It is this resource which allows an organization to pay its workers, pay off its debts, acquire raw materials, improve and expand its operations, and also to meet its day to day operational expenses.

Another important resource of management systems is information.  There are basically three main categories of information in this sense, namely, speculative, concrete, and market shared information.  Speculative information is obtained from the various media sources, where certain aspects of the organization's interests are dealt with.  The benefit of this type of information is the leads and clues it provides to the organization to delve deeper into obtaining concrete information.  However, sometimes speculative information is just not reliable and cannot be relied upon.  Concrete information, on the other hand, is based on research data or well founded conceptual theory which is useful to the organization.  In this case, the organization can access the full text of the information, study it and then decide to use it or not.  The third type of information is the market shared information.  This information is obtained from the market, for example, stocks markets, sales and financial reports publications, bank statements, tax returns, etc.

# CHAPTER SIX

## MANAGEMENT SYSTEMS TOOLS

Management systems tools are basically the instruments used to advance an organization's fortunes. Every organization has to protect its interest in various aspects of its investments, market share and capitalization, and profits. The tools of management systems are research, team work, strategic plan, and timely decisions.

### Research

Research as a tool is used by all the management systems in the organization to streamline their activities. It is this resource that allows all the systems to work in consonance with the dictates of the market and time; and as well as them working in synergy as one purposeful group.

### Team Work

A team as a tool is a functional unit of an organization, whose roles and duties are interdependent on other units in the organization, in order to achieve the primary objective of the entire organization. A team is by far a principle, tool and an activity used to enhance productivity in an organization. The efficiency and effectiveness of a team in an organization is determined by the productivity level.

### Strategic Plan

Another important tool is a strategic plan. An organization without a strategic plan will find it difficult to achieve its objectives; such an entity will be rendered impotent and will always find it difficult to deal with new challenges. The strategies could take the form of a series of decisions or a blueprint with action plans for implementation.


Technology

Also, technology is an important tool for management systems, because technology is an ingredient that mixes with other tools to ensure that management systems are appropriately and adequately applied in the organization.  Basically, technology is the methodology adopted to facilitate the application of functional roles towards the achievement of an organization's vision.

# CHAPTER SEVEN

## ORGANIZATIONAL RESTRUCTURING – A CASE STUDY APPROACH

In October 2012, a Five Year Strategic Plan report was presented to Total Supplies and Services Limited for implementation.  The successful implementation of this strategic plan, largely depend on the character, and the efficiency and effectiveness of the management systems of the company. The restructuring is therefore a boost to its internal contents, to be able to continuously implement the strategies.  The strategic plan is aimed at achieving the long term goals, mission and vision of the company.

The overall objectives for the restructuring, as a restatement are:  to achieve customer satisfaction; to become pace setters in the market; to effectively distribute quality products; to enhance team work; to earn a bigger share of the market; to restate and effectively implement strategies; to rethink and streamline its operational processes; and to significantly increase efficiency and effectiveness at an optimal cost.

There are numerous case studies and literature which points to the necessity and benefits of organizational restructuring, especially where the company is in fast transition towards growth and expansion.  The dynamism in today's volatile business environment requires all stakeholders in the organization, especially the top management to seek to adapt their entities to the kind of environment they find themselves.  An organization's environment includes suppliers, customers, competitors, and regulators.  It also includes cultural, political, technical, and economic forces.  According to Mohrman, Mohrman, Ledford, Cummings, Lawler, and Associates (1990), organizational performance is determined by an organization's character.  Specifically, an organization's performance is high when its character promotes effective exchanges with its environment and its internal design features effectively fit together and reinforce one another.

The research methodologies adopted for this process were both qualitative and quantitative data collection, with structured interview guide and questionnaire respectively. Also, the five year strategic plan report was reviewed to restate some of its key postulates.

The findings are as follows:

- The company has a five year strategic plan blueprint outlining the strategies being implemented to achieve the immediate, medium and long term goals of the company. These goals are meant to propel the company to achieving its vision or desired future state. The framework to achieving the objectives and targets constitute the key sections/themes of this finding.
- The Board of Directors remains in its advisory role, formulates policies, and negotiates contracts for the company.
- The Managing Director is the most powerful and highest decision maker in the company.
- Communication lines are vertical, running from the Managing Director down the ladder, through various work teams and vice versa.
- The Personnel system depicts a high labor turnover, where the company will have to employ more competent and specialized professionals to occupy various positions.
- The Finance system of the company depicts that, the current financial management of the company lies on the shoulders of the Managing Director. He takes the key decisions on financial mobilization and expenditure.
- The Research and Development system of the company further depicts that the company values research data to assist in critical areas of the decision making process.
- The Marketing system also depicts the need for a backup of competent sales executives and agents.

- Corporate governance is based on standardized procedures, processes, and structures led and driven by people working in teams. Leadership is driven by team work.
- On the other hand, the framework for restructuring the company was based on the following thematic areas: achieving strategic objectives and targets; management systems restructuring; business processes and operations; product and services delivery; provision of tools, equipment and standards facilities; programs and project implementation; customer services; and monopoly/dominance of market through related diversification and other solutions.
- The restructuring domain models or activities were also drawn from the following thematic areas: corporate strategy; financial planning and analysis; new business development; regulatory compliance; product management; customer/market expansion; service provision; customer satisfaction; and human resource management.

Finally, the analysis and planning of the restructuring to meet the above stated goals basically pointed to two dimensional approaches, namely: transitional approach which requires the use of centralized organizational chart; and the transition ended approach which has a decentralized organizational chart.

Total Supplies and Services limited is now a growing Ghanaian business, which sells IT, Stationery and Office Furniture products to a wide spectrum of customers. The company has a five year strategic plan with specific goal targets set to achieve, but some of the strategies cannot be implemented to the letter in the present state of the company. The company is operating in its simplistic form, whereas its operations has grown complex, which forces the company into a hunch back stature.

Therefore, in order to accomplish its vision and mission, the company has to undergo some form of restructuring to grease existing structures and also create new ones to completely and systematically deal with the challenges.

Organization restructuring is not only done for sick companies, but also for doing well companies like Total Supplies and Services Limited. The company, as already stated is expanding day-in and day-out, and its operational mass is appreciating very fast, hence the need to restructure to meet the current and future requirements of the company.

The restructuring involved the resurfacing of some key management systems, indentifying the domain models and activities and analyzing and planning the restructuring to achieve the desired results.

A growing company like Total Supplies and Services Limited needs to have well structured managements systems to drive through its programs and projects, but instead, its systems are loosely connected and applied in the simplest sense of business orientation.

The objectives for the restructuring are manifold and the key ones are as follows:

- To effectively implement strategies.
- To grease and reenergize existing structures.
- To create new systems to deal with changing trends in the business environment.
- To expand its customer/market base.
- To meet sales targets.

The challenges to be addressed from this restructuring process are as follows:

- Lack of capital due to poor access to loans, high interest rates, high taxes, etc.

- Lack of manpower.
- Inadequate facilities, like transport, space, training and repair tools and equipments.

## FRAMEWORK FOR RESTRUCTURING

Typically, doing well organizations will have an integrated management systems to enable it perform efficiently and effectively. The benchmarks set for the framework were:

a) Process
b) Products and Services
c) Tools, Equipment, and Standard facilities.
d) Customers
e) Management Systems (personnel, finance, communication, sales and marketing, research and development, etc)
f) Diversification
g) Strategies

## FRAMEWORK

| ACHIEVING STRATEGIC OBJECTIVES AND TARGETS |
| --- |

| MANAGEMENT SYSTEMS RESTRUCTURING | WORK PROCESS |
| --- | --- |

| TOOLS, EQUIPMENTS, AND STANDARD FACILITIES | PRODUCTS AND SERVICES DELIVERY |
| --- | --- |

| PROGRAMS AND PROJECTS IMPLEMENTATION | CUSTOMER SERVICES |
| --- | --- |

| ENHANCING RELATED DIVERSIFICATION; MARKET DOMINANCE AND MONOPOLY; INCREASE SALES AND PROFITABILITY. |
| --- |

(1) STRATEGIES

The company's five year strategic plan is the blueprint that defines the strategies targeting to achieve the underlining vision stated as "AFFORDABLE AND QUALITY OFFICE CONSUMABLES FOR GHANA AND BEYOND" and a mission which is also stated as "TOTAL SUPPLIES AND SERVICES LIMITED IS A COMMITTED, CUSTOMER FRIENDLY, DISCIPLINED, LOYAL AND TRUST-WORTHY COMPANY, INTO BUILDING A FORMIDABLE MARKET ENVIRONMENT FOR THE SUPPLY OF INFORMATION TECHNOLOGY PRODUCTS, STATIONERY, AND OFFICE FURNITURE AND EQUIPMENTS FOR OUR VALUED CLIENTS."

Total Supplies and Services Limited's strategies are categorized into SO, ST, WO, and WT Strategies:

- STRENGTHS AND OPPORTUNITIES (SO) STRATEGIES (GROWTH STRATEGIES)
1. Leverage good corporate image for corporate growth.
2. Leverage effective lobbying for requests for supply contracts and contract bids.
3. Leverage assets and reserves for more supplies.
4. Leverage big market share for high profits and growth.
5. Leverage more regular adverts to the availability of advertising media.
- STRENGTHS AND THREATS (ST) STRATEGIES (STABILITY STRATEGIES)
1. Leverage qualified and competent technicians and salesmen to overcome low profits and corporate decline.
2. Leverage well trained personnel to overcome poor corporate planning.
3. Leverage support to all political parties and charity institutions to promote the profile of the company.
4. Leverage effective lobbying to improve low imports and sales.
5. Leverage assets and reserves to overcome the lack of ability to supply to clients.
- WEAKNESS AND OPPORTUNITIES (WO) STRATEGIES (STABILITY STRATEGIES)
1. Deplete our liabilities with our high profits and growth.
2. Overcome internal corporate conflicts with good corporate planning.
3. Train our personnel to match corporate growth.

4. Initiate public education on peace and unity with the availability of advertising media.
5. Reduce high overhead cost with good corporate planning.

➢ WEAKNESS AND THREATS (WT) STRATEGIES (DEFENSIVE STRATEGIES)

1. Avoid high overhead cost to improve upon our profit earnings and corporate growth.
2. Resolve internal corporate conflicts to improve on poor corporate planning.
3. Reduce our liabilities to improve on low imports and sales.
4. To avoid the loss of life and property by providing regular training to personnel.

Avoid political donations to save the company from political insinuations and violence.

(2) MANAGEMENT SYSTEMS

- Board of Directors: The board of directors will remain in its advisory role, formulate policies, and negotiate contracts for the company.
- Managing Director: The position of Managing Director is the highest decision making person in the company. He takes key decisions on supply channels, sales negotiations, planning, appointments, pricing, product lines, and remuneration and rewards.
- Communication System: Communication lines are vertical, from the MD through various work teams and vice versa. The management information systems are more electronically based; but records management is by manual and electronic filing systems.
- Personnel System: The organization has to employ competent IT professionals and other qualified personnel to man its operations. The workforce might reach over fifteen personnel. The new personnel to be recruited are: General Manager; Company Secretary; Legal officer (Part-time), Finance and Administration Manager; Private Secretary/Special Assistant; Distribution Manager; Accounts Officer; Technical Manager; Operations Manager; Technicians; Cashier; Public Relations Officer; Drivers. The remuneration package shall be by monthly salary, and the

preferred age brackets are 18 – 25 and 25 – 40. The minimum and highest qualifications of personnel shall fall within Senior High School, Higher National Diploma and Professional Qualifications.

- Finance System:  Currently, the financial management of the company is centralized on the control of the Managing Director.  He takes decisions on financial mobilization and expenditure.  The Board does play an advisory role relating to financial issues.  Auditing is crucial in the financial management of the company, and external auditors are periodically tasked to audit the accounts of the company.

- Research and Development:  The company values research data to assist it make critical decisions.  It carries out market, product monitoring, consumer perception, and price index surveys.

- Marketing System:  The scope of the company's operations is beyond the northern region into upper east and west regions.  Sales are all year round, with peak seasons in January, September and October.  The peak seasons for the different products are:  January to March for IT products; April to June for Office Furniture; September to October for Stationery products.  One challenge confronting the company is lack of space.  However, this challenge will soon be dealt with when the new premises of the company become operational.

(3)  OFFICE LAYOUT

The preferred office plans or lay-outs are partitioned and landscaped offices.

(4)  CORPORATE GOVERNANCE

The company is governed according to some standardized procedures, processes and structures by people working in teams.  Leadership is driven by team work.

In order to achieve a continuous process of implementation and the realization of a changing competitive landscape, the company needs to restructure to better the five year strategic plan.  The analysis points towards the adoption of a hybrid approaches, namely centralized and decentralized approaches. The centralized approach involves a transition to the decentralized approach. The Managing Director will effectively manage a centralized operational teams or work units who will report to him directly. Then, in the long term, authority will gradually decentralized with some work units integrating the operations. This will require the provision of facilities to the different outlets of the company at different locations, to make them more business- like at an optimal cost.

As part of the restructuring process, the board should be strengthened into the highest decision making body, with the Managing Director becoming the Executive Chairman. Then a transition plan is put in place to recruit a managing director for the company, to be properly trained for the job.

The company at this stage requires highly trained and motivated personnel to work as a team to achieve the goals of the company. Basically, the restructuring will ensure high level planning, critical thinking and analysis process, and an excellent market transition into an assembling plant, wholesaler, retailer, and based on an efficient distribution network.

# RESTRUCTURED ORGANIZATIONAL CHART (TRANSITION CHART)

## CENTRALIZED APPROACH

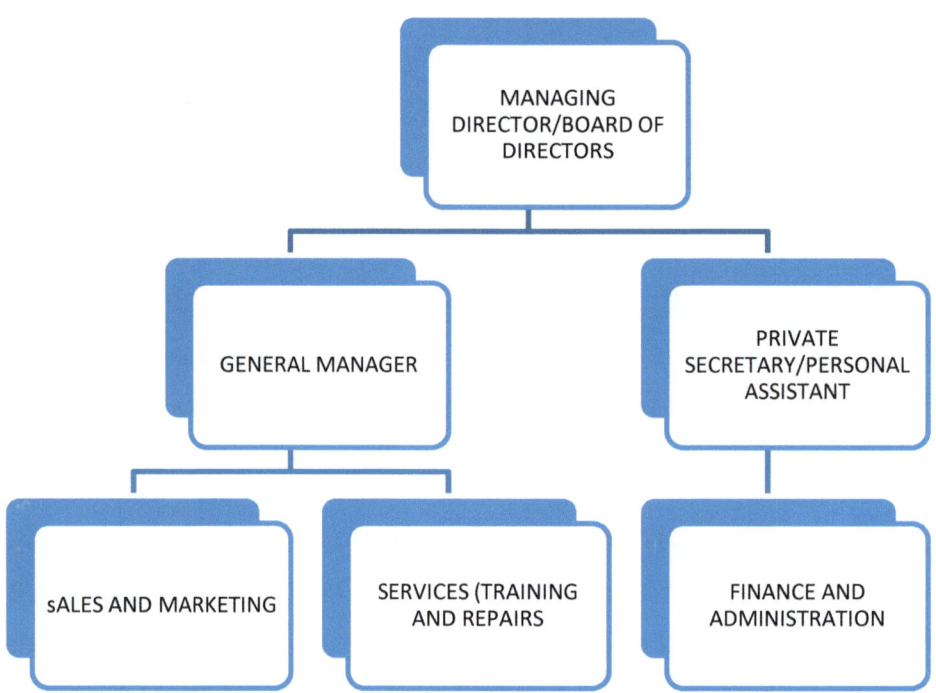

The Managing Director is the central authority; the BOD advises; the General Manager and Private Secretary/Personal Assistant are his two wingers and they do what is delegated to him; the other departments reports directly to him or per the General Manager and Special Assistant.

# RESTRUCTURED ORGANIZATIONAL CHART (PERMANENT CHART)

## DECENTRALIZED APPROACH

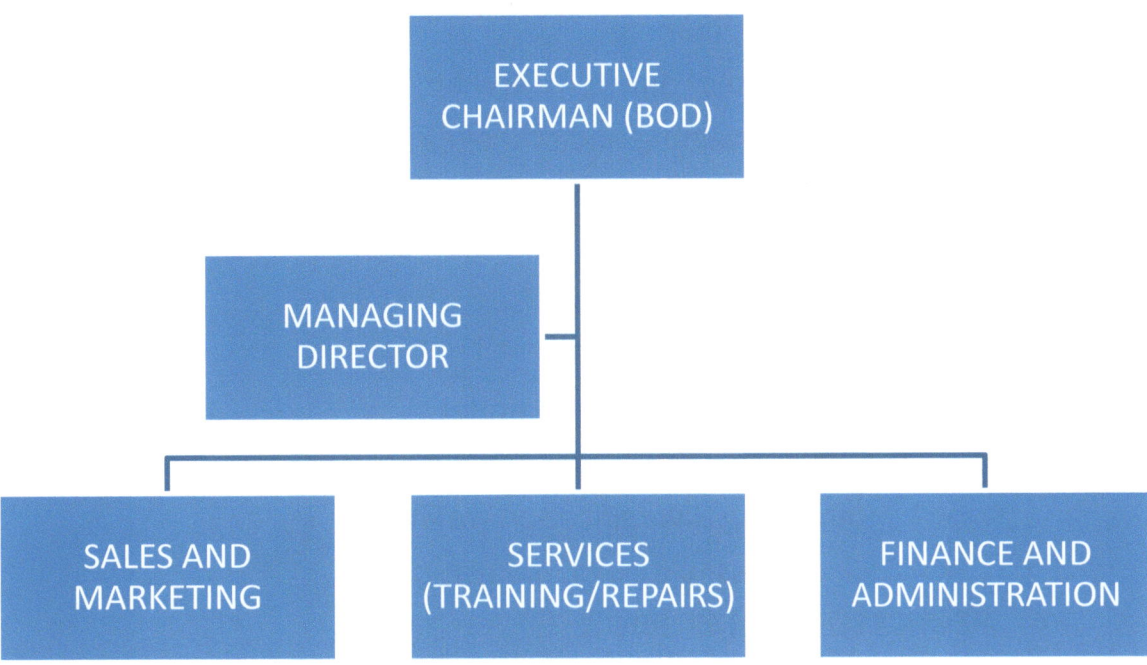

The chairman of the Board of Directors will act as the head of the company, and directly participate and supervise the Managing Director in the day-to-day running of the company. The managing director will consult him before making certain key decisions, which will be spelt out in his job description. However, the Managing Director shall hold the responsibility of the smooth running of the business or otherwise, and shall be held responsibilities for misapplication or management of the company, as far as his jurisdiction lies.

The key activities for the restructuring are spelt out below in nine key areas of activities:

1) CORPORATE STRATEGY
   - Achieving corporate vision.
   - Fulfilling corporate mission.
   - Achieving corporate objectives, goals and targets.

- Efficient and effective corporate management.
- Efficient use of profits.

2) FINANCIAL PLANNING AND ANALYSIS
- Long term planning
- Preparation of quarterly and annual financial statements.
- Effective planning and management of resources.
- Efficient and effective capital management.
- Efficient application of profits.

3) NEW BUSINESS DEVELOPMENT
- Achieving sales targets.
- Intensifying market techniques.
- Capitalizing on tender bidding opportunities.
- Intensifying diversification (related).
- Explore possibilities of acquisitions.

4) REGULATORY COMPLIANCE
- Strict compliance to tax obligations.
- Regular auditing.
- Seeking guidance on regulations relating to tax, procurement, trade, banking, imports and exports, etc.

5) PRODUCT MANAGEMENT
- Develop new product lines within the three main products the company is selling, namely IT, Stationery and Office Furniture.
- Regular stocks and inventory of goods.

6) CUSTOMER/MARKET EXPANSION
- Intensify sales.
- Seek orders.
- Ensure prompt delivery on orders.
- Ensure good customer relationship.
- Fulfill commitments.

7) SERVICE PROVISION
- Provision of facilities, tools and equipment.

- Recruitment of qualified trainers/repairers.
- Provision of training manuals.

8) CUSTOMER SATISFACTION
- Provision of customer hotline.
- Easy trouble reporting and resolution platform.
- Technical/Product support.
- Tracking reliable customers for special packages.
- Maintaining reliable customer account.

9) HUMAN RESOURCES (PERSONNEL)
- Recruit talents.
- Compensation/reward packages.
- Performance bonus.
- Professional development and training.
- Employee relations.
- Compliance to company code and labor laws.

# CHAPTER EIGHT

## PUBLISHED ARTICLES

Article One

APPLYING MANAGEMENT SYSTEMS IN THE AGRICULTURE SECTOR IN GHANA

Management is both a theoretical and practical activity involving the science of planning, organizing, staffing, directing, coordinating, communicating, controlling and decision-making.  These activities are managerial processes which are apparently necessary and systematically applied.  Management systems can be defined as the main structures or pillars upon which an organization is built, and which is apparently necessary for the successful realization of management goals and targets.

The application of management systems in any organization is very important for two main reasons, namely, resources utilization and the management of risk and profits.  All organizations, according to Howard Barnett (1992) in his book "Operations Management", have only five different kinds of resources to work with, which are equipment, material, money, people and time.  The basic idea of resource management involves the adoption and application of these management systems.  Secondly, in modern business organization, where the private sector has a role and ownership rights, the management of risk and profit is the revolving factors that critically needs the application of management systems.

Statement of Problem

The management systems in the agriculture sector in Ghana pose some acute challenges.  Food is a basic necessity of life on which man depends for energy, but unfortunately, the domestic production level of the major food crops is very low to suffice the enormous population.  This has necessitated the importation of food products from abroad.  The agriculture sector has continually faltered in productivity due to many factors, including ill established or no management systems.  Statistical figures indicate that the sector is dominated by small scale farmers who lack adequate resources and have poor background in risk management. Another challenge is the level of illiteracy and non application of management systems. Storey (1982) shares the findings of a research which showed that, firms established by graduates performed significantly better, in terms of turnover, than otherwise similar firms established by non graduates. He also quotes another research which suggested that, the nature of education is one factor which distinguishes the craftsman type entrepreneur from the opportunist. Thus, for a successful business organization, the correct application of managerial expertise and roles are necessary.

Therefore, my main and first objective for this article is to amplify the relevance and necessity of the application of management systems in modern business organization. This suggestion may look vague to the majority of peasant farmers and households whose object for production is to feed their mouths, but not as a matter of profit. On the other hand, one can argue proactively that, these farmers can still be organized into cooperatives and groups through which their resources are harnessed to make the lot out of them. Secondly, it is to build the capacity of our farmers in management systems, to enable them appreciate the necessity to adopt them for the attainment of their goals and targets. Thirdly, this is likely to provoke more interest in the agriculture sector by young entrepreneurs and experts alike, and may bring on board more useful ideas and inputs to be shared by all stakeholders in the sector.

Management Systems

The celebrated management theorist, Henry Fayol defines management as, to manage is to forecast, and play to organize, to command, to coordinate and to control. These are classified as elements or functions of management, but these functions cannot be performed in the vacuum, instead they are applied through the management systems or structures. These systems include: personnel, production, marketing, finance, communication, research and development, and other systems found to be necessary. Etienne et al (1992) defines a system as a set of components linked by relatively organized relationships, in-order to fulfill certain functions. Management systems does not discriminate the levels of management, instead, it collectively coordinates the activities of all the different levels of management into one common front, for the achievement of the organizational goals and targets.

Personnel System

The personnel system in the agriculture sector involves two important levels of human resources. The first is the individuals or groups or people who have taken risk to venture into farming as a business, and are termed peasant, small scale or commercial farmers. The second level are those trained and employed by the government to provide extension services and other expertise to farmers in the fields and as well as others operating within the NGO fraternity and whose efforts are geared towards improving the incomes and livelihoods of these small scale farmers. The personnel system is important for the following reasons: the employment and manpower development the sector provides for the people; secondly, the education and training the people receive to build their capacities; thirdly, the industrial relations, health and safety factors that are addressed; and most importantly their incomes and remunerations. Hence, there is the need for a careful implementation of a workable human resource strategy in the sector.

Production System

According to Etienne et al (1992), the production system in agriculture perspective is characterized by a certain type of combination of production means like manpower, land, equipment, in order to produce crops and to meet the farming requirement of food self sufficiency, taxes, savings, etc. He further concludes that, the evaluation criteria of the organization will largely depend on the effectiveness, efficiency, impact, and viability of the production system. The production system is crucial not only for its crop yields, but the additional functions it plays, like product quality, packaging, standardization, simplification, and specialization. All these functions ensure quality products and limited varieties of products, which guarantees easy market access.

Marketing System

Marketing system is an entrepreneurial task involving the judicious use of limited resources for the speedy disposal of end products to the consuming public, with the aim of making profit. According to the British Institute of Marketing, marketing is a management process responsible for identifying, anticipating, and satisfying consumer requirements profitably. The function of the marketing system includes product planning and development, distribution, promotion, and pricing. The efficient direction of the marketing system is the key challenge of the agriculture sector, and it is that which accounts for much of post harvest losses for decades in the country.

Financial System

The financial system of any organization is the actual generation and management of scarce resources by the entrepreneur or investor. There must be available funds classified as the initial investment capital to purchase land, equipment, material, working or revolving capital for overhead cost and expenses, salaries and wages, etc. The main function of the financial system is to manage risk to ensure maximum use of scarce resources for high profits.

Communication System

Mardock and Scutt (1993) defines communication system as the source which passes information and instructions which enables a company or any other employing organization to function efficiently and effectively, and employees to be properly informed about developments. Communication also involves the transmission of information to other stakeholders of the organization, such as customers, the general public and other relations and as well as the total administrative processes. Thus, it embraces the organization's management information systems (MIS), administration, records management, and computer based systems.

Research and Development System

Research and development system is an important pillar in organization theory, because it is dependent on statistical data. The main functions of this department is initiating and introducing, improving and updating, maintaining and innovating the other systems or departments, such as production, marketing, etc, by using approved statistical data. Marketing research will involve demand/consumers, competitors, product, sales, distribution, promotion, etc. Research in personnel system will involve work, methodology, efficiency and effectiveness, etc. Research in production will also involve quality assurance, value addition, product development, etc.

Other Management Systems

The management process is dynamic, and every situation can be dealt with through restructuring of the management systems, especially with the case of developing countries like Ghana, which are said to have their peculiar challenges which is impacted upon by our culture and history. Thus, the application of other management systems to deal with certain special initiatives, such as the presidential special initiatives is possible and can deliver the desired results. Sometimes, some of the special initiatives may target, certain types of commodities aimed at developing and improving their value for export and for domestic consumption as well.

Statement of Analysis and Conclusions

The current situation in the agriculture sector relating to management systems can be described as follows: The personnel system is characterized by well skilled personnel working under the ambit of the Ministry of Food and Agriculture and its agencies; whereas predominantly illiterate personnel undertake farming as a business for their livelihoods. The production system is, on the other hand characterized with high risk due to unpredictable weather patterns, poor irrigation systems, lack of modern farming skills and equipments, etc. Productivity is low and majority of farmers are only effective during the raining season. The market system is no different, due to poor product quality and value, and inadequate industrial and trade support from the government and other stakeholders in the business. The financial system is very appalling, very poor and farmers lack the necessary capital to invest in their business. The communication system is hampered by inadequate opportunities to build the capacities of agro-based entrepreneurs. The research and development system is also mal functioning because the farmers do not have the means to acquire the know-how to access new technologies and inventions.

The remedies are that, the government should entice the highly skilled personnel to take up agriculture as private business, instead employing them under the Ministry to provide only technical services to farmers.  Possibly they can replace the current crop of illiterate and small scale farmers who hardly make ends meet in the sector.  Secondly, farmers should be assisted to upgrade the quality of their products to meet the international standards for export and domestic consumption.  They should be encouraged to specialize in the crops they cultivate, in order to cease from being 'the jack of all crops and master of none'. Adequate financing through trade and low interest loans, will go a long way to improve the agriculture sector.  Government should grant more subsidies to farmers to enable them embrace modern farming methods.  Lastly but not least, the Ministry of Food and Agriculture should devote its budget on research and development than it is doing now, to develop products and value analysis regimes that will empower farmers to explore and access opportunities in the sector.


Article Two

MANAGING THE RICE INDUSTRY IN GHANA FOR SUSTAINABLE FOOD SECURITY AND ECONOMIC GROWTH

As far back as the seventeenth and eighteenth centuries, rice had already become a commercial food crop in Ghana (Jean, 1994).  In 1996, one interviewee in a survey on the rice industry stated that, as far back as 1911, MoFA had started testing the suitability of rice varieties such as 'America', 'Pindi', 'Bindombi', and others.  The production of rice became widespread when the Ghanaian-German Agricultural Development Project was established in 1974.  Under the project, larger expanses of rice fields were cultivated, with an average of 5,500 hectares annually; and this continued until 1978.  Also in the 1970s, rice production benefited from big government subsidies, which resulted in the formation of the

Ghana Rice Growers Association in 1972, thereby boosting organized rice farming in the country.

Under the Irrigation Development Authority, some irrigation projects were undertaken in the Northern region; the Botanga irrigation project had 1,200 hectares, Libga 40 hectares and Golinga 65 hectares respectively. In the Upper East region some other schemes at Tono and Via benefited some of the operators in the industry.

The increase in the number of rice farmers influenced the Seed Division of MoFA to increase their stocks of rice seeds from 8,000 metric tons to 10,000 metric tons between 1995 and 1997 crop seasons. The rice varieties distributed at that time were GR 18, Mendi and Faro 15.

Primary Production of Rice

The single most important factor of production for rice is land; this has made the consideration for land very critical in the industry. Statistics show that, domestic production of paddy rice has steadily increased as a result of the availability of arable land. In 1981, the land area for rice cultivation was 116,000 hectares and yielded 97,600 metric tons of paddy rice; whereas in 1990, 80,000 hectares were cultivated and yielded 132000 metric tons. Also, in 1993, 157,000 metric tons were produced; and recently in 2011 domestic production of rice has increased to 463,000 metric tons.

Paddy Rice ready for parboiling at the Lolandi Rice Processing Center

## Consumer Behavior

Our primary objective is to produce, eat and sell the surplus, either domestically or export. In either case, marketing is critical for producers. However, the bare fact about the local rice is that, consumers prefer imported or polished rice to the locally processed rice. Incidentally, we continue to pretend that, the local production of rice is far below what is demanded for, thereby creating the compelling need to import more from elsewhere. The question is whether we are actually importing to augment a shortfall in rice production in the country. What is true is that, there is a general shortfall in the production of other local staples which creates the necessity to import rice. In a country report by MoFA in 1991, it was stated that, agricultural production have not been able to keep up with the growing demand for food, and part of the demand has been met by increased food imports.

## Research

The question is how much research has been done in the production of rice, and how many of the recommendations have been implemented. Mboob, (FAO, 1996), lamented that, scientists spend many years doing research which they hope will improve agriculture production and are then disappointed to find that farmers take no notice of the recommendations. There is actually an in-balance between the level of research done on rice and the rate of productivity in the industry. Though we have seen a steady rise in the production of paddy rice, it is still inadequate compared to recent technological development arising out of research.

The West African Rice Development Association (WARDA) in collaboration with FAO and ECA has done a lot of research in the past, and has disseminated much information on rice production. The Savannah Agricultural Research Institute (SARI) is no exception. An interviewee (1996) spelled out a number of research conducted for the development of rice production in Ghana, which included; breeding and selection, weed management and control, relay cropping and cropping systems, fertilizer application, characterization of rice valleys, etc.

## Application of Modern Technologies

Modern machinery and equipments, such as tractors, planters, combine harvesters, dryers are used today by some commercial farmers. An inventory of four wheel tractors in Ghana in 1994 by MoFA showed that, the Northern region had the largest number of them; there were 875 unserviceable tractors, and 1,637 serviceable ones, totaling 2,512 tractors. These figures should be doubled if not tripled over the past nineteen years. Other modern techniques devised for rice production are; bonding, irrigation, transplanting, weed control, pest control, improved seed varieties, harvesting processes, and storage methods.

Private Sector Participation

The rice industry in Ghana is more or less an aggregation of small scale rice farmers, which is the basic characteristic of most small scale industries in developing countries. Ntim (1986) wrote that, there are some obstacles that entrepreneurs in small scale industries must face in developing countries. These obstacles are: lack of basic technological knowledge; lack of purchasing power and capital to establish and operate business; inability to implement research findings; insufficient facilities for repairs and maintenance of equipments; inadequate facilities for technological training; scarcity of basic entrepreneurial knowledge and skills of economic factors in business operational management; and finally, insufficient information about available local materials and technologies used successfully elsewhere. The private sector in the rice industry is facing a mired of problems, which in my opinion are similar to those stated above.

Local Rice being processed and graded manually at the Lolandi Rice Processing Center

The Role of Agricultural Development Bank (ADB)

In the 1960s the government noted with concern the bias against advancing credit to the agriculture sector and rural people. To remedy the situation, the National Investment Bank (NIB) was established in 1963 to provide credit to both industrialists and large scale farmers. Later in 1965, the Agricultural Development Bank (ADB) was established to provide credit primarily to small scale farmers for agricultural development in Ghana (Owusu 1986). In the light of this mandate, the bank could do more to revamp the rice subsector, since majority of the operators are small scale farmers. The bank could also evolve a strategic investment portfolio for the sector, which might result in more capital investments from potential strategic investors, since rice cultivation requires the injection of more capital. In any case, the potentials of the small scale farmers, who normally form themselves into groups to qualify for small credit, should not also be taken for granted. They need a lot of financial support if we are committed to raising the standards of living of our rural people.

However, just as the financial institutions need to do more to assist the farmers, so is the government too, has a responsibility to look into addressing these teething problems facing not only the rice industry but most of the other subsectors in the agriculture sector. These problems include: rising cost of inputs like fertilizer, chemicals, etc; rising cost of services of tractors, combine harvesters; stiff competition from imported rice; insufficient credit to rice farmers; vagrant weather conditions; and the menace of post harvest losses related to storage, bushfires, etc.

The Importance of Management Systems

Management systems is important in the industry because of the solutions it can provide to the mired of problems bedeviling it. The specific areas of the rice industry management systems will offer important remedies include: mechanization; marketing management; planning and international marketing; action methods, planning, and taking; managing time; profit planning; motivation; and field management.

Agricultural mechanization is a big problem in the industry, due to the inability of farmers to purchase machinery, inadequate maintenance and lack of spare parts. Secondly, marketing management and planning is critical to overcome competition from imported rice. Thirdly, decision making will be facilitated for other gains to be made. Also, time management is essential to ensure effectiveness and efficiency in the coordination of activities of the stakeholders in the industry. Time management can further be emphasized for the unpredictability of the weather, which calls for constant monitoring of both dramatic and marginal changes in weather patterns. In addition, the need for profit planning should be inculcated into the mindsets of our farmers, to change their attitudes towards looking more business-like. They must be reoriented to see themselves as entrepreneurs and profit earners. Time management is still mentioned because, it transcends into field management, which embraces timely cultivation, irrigation, sourcing and application of inputs and the meticulous care given to the crops.

Food Security and Sustainable Economic Growth

For the achievement of food security and sustainable economic growth in the industry, we need to understand the dynamics of the industry, so as to improve those aspects that require it. In a survey conducted in 1996, the report showed the following demographic outlines of the industry, though some changes might slightly happen, but the picture is very likely to remain pretty the same, as the change process in our system is by far and large relatively slow.

The rice industry is predominantly private small scale and commercial farmers. Rice is cultivated both as a staple and cash crop. About 87.5% of the farmers cultivate rice for sale and 12.5% for household consumption. Illiteracy rate is 35%, farmers with non-formal education 12.5%, and those with formal education, either primary, junior high school, senior high school or university education 52.5%.

About 28.75% of rice farmers cultivate between 1-10 acres; 31.25% cultivate 10-50 acres; 18.75% between 50-100 acres; and 18.75% cultivate 100 acres and above. For sustainable food security and economic growth, these figures needs to be improved upon and the appropriate linkages established, to ensure the successful implementation of strategies for achieving the desired goals.

Conclusions

The rice subsector has the potential to provide the sustainable food security and economic growth we so much desire, if the following conditions can be improved in the industry.

i)    The capacity of enterprises and entrepreneurs in the industry, especially the knowledge and skills of farmers.
ii)   Increasing the number of acreages of farm lands and soil fertility.
iii)  Accessibility to ready capital, inputs and other logistics.
iv)   Provision of irrigation facilities for both commercial and small scale farming communities.
v)    Effective disease and pest control mechanisms.
vi)   Access to improved seed varieties based on industrial specifications.

vii) For purposes of creating, expanding and revamping the industry, to meet the two basic objectives of having ready local and international markets, and jobs for the massing youth, it is necessary to restructure the entire agriculture sector. The frame work for doing this, in my opinion, could be by stratifying the agriculture sector into crop and livestock zones around the country. This strategic management policy should be a model with a medium to long term planned objectives, coming with it an effective monitoring and evaluation team which will embody key stakeholders in the industry. The plan calls for specialization in cropping and livestock rearing based on each zone's comparative resource template. For example, each zone specializes in two crops/animals production on a large scale, so that zones which do not produce the products they need will buy them from other zones. This will encourage local market and job creation. The situation we have in our hands today, is more or less 'jack of all trade' sort of business, whereby people merely produce what they need but not what others need, which does not in itself encourage buying and selling among the people. Money is difficult to come by because business is too low among the people. Post harvest losses related to lack of market could be a thing of the past.

viii) Adequate and durable storage facilities will to a large extent, reduce post harvest losses and pricing.

ix) Quality assurance for the product for both local and export market, through the establishment of ultra modern rice milling factories and other applied techniques.

Article Three

## RURAL URBAN MIGRATION VERSUS RURAL INDUSTRIALIZATION IN GHANA AND AFRICA

### Introduction

Rural Urban migration is a menace with different faces; sometimes the migrants are compelled to move, in other situations they move on their own volition.

Whichever way it takes, the consequences are the same. These migrants always presume that, there are greener pastures or viable opportunities in the cities than in the rural areas, which they are going to meet. Unfortunately, to their dismay, upon arrival, they are faced with the worst hardships and conditions of their lifetime. They even sometimes will prefer to go back to where they came from, but if not for the "red marks" some of them left behind in their villages.

Many and varied factors can be attributed to rural urban migration, which includes: lack of social amenities like schools, hospitals, entertainment centers, etc; lack of economic ventures except farming, which is less attractive to the youth; break down of extended family systems which compels the vulnerable to flee from persistent hardships; lack of public and private sector businesses to create employment opportunities; the influence of the mass media; and lack of industries to create connectivity between the rural folks and urban dwellers, and means of employment to ease the burden of boredom and persistent poverty syndrome.

For the sake of those who may argue otherwise to the factors influencing rural urban migration, let me explain the points stated above, before tackling the reasons why rural industrialization can serve as the antidote for it.

Lack of Social Amenities

Society is supposed to be dynamic over time, but if the ways things are done remains static, then the two likely scenarios will be desertion or conflict engrossing the society. Every human organization needs growth and development to take care of the needs of its members, at any given time. Otherwise, the people living within it will find it not worth living in, since the entity will be characterized with numerous challenges. Lack of educational institutions in the rural areas is one important reason why many youth leave their villages to the urban centers. Despite the importance of education in the rural areas, it will be difficult to site a university or any other higher institution in a rural area, because education is primarily about accessibility and greater numbers of the populace live in the urban centers, thus, the need to build higher institutions in the urban centers. Poor roads are a disincentive to rural life, and may cause many rural folks to migrate to the urban centers. Due to the poor nature of the roads, fewer goods and services can be transported into the hinterlands, and the people will be compelled to travel to the cities themselves to look for those goods and services themselves. Similarly, their farm produce cannot be transported to the market centers, thereby causing them to remain in abject poverty. Also, lack of primary and tertiary health care facilities;

entertainment centers, etc, can force the rural people to migrate to the urban centers, to the detriment of their families whom they should have taken care of back in the villages.

Lack of Economic Ventures

Most families in the rural areas are engaged only in farming, which is seasonal. About half of the year is wasted in idling, performing funerals, playing 'oware', or sitting by the road sides to watch vehicles that pass. In order to evade the cycle of waste and boredom, the youth find the rural areas very boring to stay. Therefore, those of them who leave their home villages to further their education do not come back upon completion, because their services are no more needed under the circumstances. In this respect, their inability to return back to the rural areas is because, their educational attainments makes the places unfit for them. Those places lack the requisite businesses and vacancies to absorb personnel of their caliber.

The Break Down of Extended Family Systems

In the past, extended family systems were working in many societies across Ghana and beyond, especially in the Northern parts of Ghana with some traditional norms and practices still being observed in their 'primitive' forms. Under the extended family system, being a family head mandates you to take good care of every family member, be he an orphan or not, or delegate the responsibility to another member. A young man is trained in the rudiments of farming and finally he is found a wife to marry to enable him begin his own family life. If a young girl, she's betrothed to a man who takes up her responsibilities, otherwise, she is handed over to her auntie, who trains her in the rudiments of womanhood until she wins someone's heart and marries him to start a new life. This time, these traditional values and practices are outmoded, and because of the school system and the responsibility that come with it, when your immediate family is lacking the capacity to fend for you, then it becomes a hell of a problem to the poor head. The emergence of the 'Kayaye' menace in Accra, Kumasi and other big cities around the country is a classical example of what I am pointing a finger at.

Lack of Public and Private Businesses

Most government and private businesses are only concentrated in the urban centers, where majority of the people living there need their services. Those in the rural areas, when they need these services, will have to travel long distances to reach them. However, if some of these businesses could be established in the rural areas as well, at least their outlets, such as clinics for health services, sales and marketing outlets for whole sale and retail businesses, local government councils, transport terminals, and many others, the people could be motivated to stay back to improve their own lives. As most of them travel to the cities frequently to access some of these services at higher costs, their exposure and experience can influence them to see the rural areas less attractive.

The Influence of the Mass Media

With the advent of new innovations and technologies in the way we go about our lives, many youth in the rural areas are beginning to question the cycle of simplistic lifestyles in the rural areas. This problem is compounded by the proliferation of the mass media. Some of them buy TV sets, video and radio sets to learn about other people, and then make comparisons, resulting in their dissatisfaction of their current state of affairs. The consequence of this complex

feeling and misconception in their state of minds is the decision to leave to the cities where they can enjoy better facilities than they have in their home villages.

Lack of Industries to Create Connectivity between Rural Dwellers and Urban People

The scenario is that there is little or no connection with our gallant farmers and local industries for the supply of raw materials. Most of the factories and farmers complain every now and then over issues relating to lack of adequate supply of raw materials or lack of adequate market for the farmers produce. Some companies organize the farmers into small holder groups or out grower schemes in a bit to secure a certain quantum of supply annually from the farmers. In such cases, sometimes the farmer's interest is not well taken into consideration, so these schemes do not see the light of the day. Another dimension to it is the lack of industries that are capable of consuming the raw materials produced by farmers annually, to create a sustainable means of livelihood for our gallant farmers.

Remedies to Rural Urban Migration and the Hypothesis of Rural Industrialization

Having discussed the factors influencing rural urban migration, it is ripe to find remedies for it, and this is where rural industrializations fit in as a remedy. Fosu (1991) writes that, agriculture accounts for about 51 percent of the economy, whereas the manufacturing sector takes about 5 percent and the services sector 4 percent respectively. Even though these figures might change now, but there is still the basis to argue for or against certain trends of affairs in Ghana and Africa in general. Since agriculture accounts for over 50 percent of the Gross National Product, it is sufficient to say that, our industries should be more agro-based than other products. Raw materials will be cheaper to find and cost of production will therefore fall, and logically go a long way to enrich the poor farmers.

These agro-based industries should be established at places where the raw material is most produced. For example, given the scenario that a manufacturing company called BAF Company, processes cocoa beans into chocolate is Established, in Bibiani where cocoa production is their main occupation; this factory is very likely to bring numerous opportunities to the people living in the district, such as ready market for their produce, source of employment for the youth, roads will be constructed, and connection will be made with their counterparts in the cities through trade and the media. Another way to boost rural industrialization is to mechanize agriculture and apply modern technologies that will keep farming activities going on throughout the year, but not as a seasonal affair. In addition, the gaps explained above should be closed up concurrently, alongside the industrialization agenda. This means that, the agriculture sector will employ less numbers of people, than are employed today. For example, the Nsawan Food Canary and Palgu Tomatoes Factory are prime examples of what I am hammering on. These factories failed in the past, not because the scenario planning was not feasible, but rather because of man-made factors. For every enterprise to succeed, human attitudes must also change. The second reason is that, the farmers will redeploy to more lucrative ventures, even if it entails migrating to other places like the cities.

Also, for these industries to succeed, then the issue of how they will market their processed products should be tackled simultaneously. It is one important factor, which probably caused the failure of the previous ones in the country. The attitudes of Ghanaians must change to patronize their own products, coupled with the production of quality products on the side of the manufacturers. Otherwise, if our industries are to target only the export market, then we are likely to face the same fate as before. A systematic drive to change our Ghanaian tastes and preferences must be undertaken, with the support of all stakeholders in our society. The scenario is like, National Development Planning should not only be dependent on strategies, but rather, more on the will of the people to make sacrifices to implement and work towards the achievement of the country's broad objectives set out in the plan. That is why over the years, many of our development plans, if there has been any, have gone through the drains. In our current state of nation building, dogmatism is the ideal watch word for all. The recent 'Friday Wear' slogan being pursued by the government and accepted by many Ghanaians is in the right direction.

There may be other factors for which some people may not agree with the above stipulated remedies for rural urban migration, but it is my view that, everybody understands that the phenomenon is dangerous and a hindrance to national development. This convergence of view-points is what is crucial to finding a lasting remedy to the menace. In this article, my effort is on how Ghanaians and other people within and beyond the continent who face this kind of problem can resolve it. My hypothesis is the proposal that, rural urban migration can be resolved through rural industrialization, which in itself requires a certain process of collaborations and precautions. Rural Industrialization may not be the sole remedy to the menace of rural urban migration, but at least it can drastically reduce the menace if properly planned well.

Current Scenario in Ghana and else where

Unfortunately, what we see most often is, the factories are located and concentrated in cities like Tema, Accra, Kumasi, and Takoradi, etc, where both skilled and unskilled labor force is concentrated, to the disadvantage of the rural areas. Thus, leads to population explosion and the emergence of urban slums with high rate of malnutrition, disease and crime. The main feature of the rural areas is thus, to find many of the active population engaged in farming, which goes no further than subsistence agriculture. According to the World Book Encyclopedia (Volume 1), millions of farm families in Africa, Asia, and Latin America produce barely enough food to subsist on, that is, to meet their own needs, whereas the great majority of farms in the industrialized world are commercial farms. If we could do more in Ghana and Africa by mechanizing agriculture and applying modern farming techniques like irrigation and farm warehousing, coupled with the establishment of agro processing factories, to process the raw materials into finished goods, then, the value of our export commodities will appreciate to the satisfaction of everyone in the sector. Another advantage of it is that, food will be cheap and available, and our living standards will also appreciate because, we will be engaged in other viable businesses that will empower us financially to meet other essential needs.

Conclusion

When people are convinced about concrete evidence of progress and plans for the future, then they will very much believe they are likely to see the light of day through the tunnel. I strongly believe that, nation building is achieved through a systematic development of both the urban and rural areas. A disconcerted effort can lead to an in-balance in the distribution of national income and infrastructural action plans, such as education, health, democracy, etc.

In this article, my proposal to resolving rural urban migration and ensuring the equitable distribution of income and infrastructure in the country is through rural industrialization. On the other hand, I do not discount the concurrent resolution of the other factors militating against rural development. Firstly, the adequate provision of social amenities in the rural areas; secondly, the establishment of economic ventures that employ the idle hands, and most especially the youth; the extension of social and welfare services to rural areas to assist needy families and individuals; and the establishment of local government and private businesses. With these measures, we can reduce the percentage of our population engaged in the agricultural sector, and will still remain self sufficient in food. According to the World Book Encyclopedia (Volume 1) about 60 percent of the work force in Africa and Asia are farmers, whereas only 2 percent of Americans and Canadians farm.

In terms of productivity, Europe, USA and Canada put together amount for 41 percent of the food production of the world, though only 3 percent of their population are farmers. This shows that, methodology, policy, and technology are more important in determining the country's productivity than its labor force engaged in the sector.

Foster (1982) in his book "Mastering Marketing" writes that, impact analysis is a version of scenario writing, it concentrates on the impact various forecasted technological development will make on a particular industry. The scenarios therefore given in this article are meant to determine the length of success rural industrialization will bring to the vast majority of rural people.

An issue which is not yet dealt with in this article, but equally important is the land tenure system prevailing in the country. According to Rufai in an article published in 1988, wrote that in the Tamale areas, as far back as 1987, it was

found that only 14 settlers indicated they had given to land owners kola nuts, the rest of the respondents did not mention any form of payment for land acquisition. Though the dynamics of land tenure system may change today in Ghana, including the northern regions, but land needed for viable projects and investments should not be difficult to find.

Lastly, but not least, is the issue of climate change which has bedeviled mankind for the past decade, but it is arguable that, when the industries are not concentrated in a particular area, unlike is the case in Tema, then the rate of population is minimal. Moreover, agro based industries' emissions are less toxic than chemical, plastic and metallic industries. More so, these industries to be established can be planned to address all issues of the environment.

# BIBLIOGRAPHY

1. Barnett Howard. <u>Operations Management.</u> London, The Macmillan Press Ltd, 1992. Chapter 1 to 16, Pages 1 to 394.
2. Storey D.J. <u>Entrepreneurship and the New Firm.</u> London, Croom Helm Ltd, 1982.
3. Beaudoux Etienne, Combrugghe Geneviere De, Douxchamps Francis, Gueneau Marie Christie. <u>Supporting Development Action.</u> The Macmillan Press Ltd, 1992.
4. Muhlmann Alan, Oakland John, Lockyer Keith. <u>Production and Operations Management.</u> London, Pitman Publishing, 1992. Chapter 3.
5. Mills Geoffrey, Standingford Oliver, Appleby Robert C. <u>Modern Office Management.</u> London, Pitman Publishing, 1978.
6. Forster Douglas. <u>Mastering Marketing.</u> London, Macmillan Press Ltd, 1982.
7. Dobbins Richard, Witt Stephen F. <u>Portfolio Theory and Investment Management.</u> Oxford, Martin Robertson & Co Ltd, 1983.
8. Scutt Carol, Murdock Alexander. <u>Personal Effectiveness.</u> Oxford, Butterworth Heinemann Ltd, 1993.
9. Bozza Jean. <u>Development of Rice Production in Northern Ghana.</u> CIRAD NEAS, 1994.
10. Bozza Jean. <u>Rice Development Project in Northern Ghana.</u> CIRAD NEAS, 1995.
11. Fayol H. <u>General Industrial Management.</u> and <u>Organization Theory, Selected Readings.</u> Edited by D.S.Pugh, Part Two, Chapter 9, Page 135.